GLASGOW,
SCOTLAND

MOROGORO,
TANZANIA

To my grandchildren:

May you find ways to use your talents to make a difference in the world.

In gratitude to Dr. Donna Kean for sharing her inspiring story and to Lily Shallom for helping me better understand the unique and wonderful work of APOPO.

The author is donating a part of her royalties to APOPO.

ISBN: 979-8-9914982-2-7 (hc)   Library of Congress Control Number: 2024927412

979-8-9914982-3-4 (pb)   Library of Congress Control Number: 2025927746

Cover and layout created by Tanya Space.
Copy edited by Ronda Roaring.
Illustrations on the end sheets were created by Sadie Peers.
All other image credits are referenced in the back of the book.
Fonts used for this book are Adobe Garamond Pro, Avaline Script SC Sketch, Corporative Sans Rd and Rock Salt Pro Regular.

This publication includes information from the scientist and the organization from 2022-2024. Scientific information and practices undergo updates and revision, therefore some of the information in this book may not reflect current procedures being used at this time.

First edition, 2025

RATS
WITH BACKPACKS

Dr. Donna Kean and Her Work at APOPO

J. N. Courtney

"Good actions give strength to ourselves and inspire good actions in others." -Plato

APOPO: A Dutch acronym for Anti-Persoonsmijnen Ontmijnende Product Ontwikkeling. In English, it translates to Anti-Personnel Landmines Detection Product Development. A global non-profit and Belgian NGO (non-governmental organization), it researches and develops scent detection applications for African giant pouched rats and dogs, to save human lives and improve the environment.

J. N. Courtney Publications; Penfield / New York.

DONNA

ZACK

Donna Kean's love of animals began with a scruffy wild dog named Zack.

All of the neighborhood kids knew Zack, but he didn't seem to belong to anyone. One day, after Zack had been in a car accident and his paw had been injured, Donna's dad gently carried Zack into their family flat.

This was the first time that Donna would observe and help to tenderly train an animal. At first, Zack didn't like being clean or stuck inside the flat. He was used to roaming the streets and would try to sneak out the front door whenever he could. But he was smart and learned quickly how to be a family pet. Donna grew to love Zack dearly—her first animal friend—especially when he would curl up beside her when she wasn't feeling well or was unhappy about something.

Donna was born in Glasgow, Scotland on November 8, 1988. Although she grew up in a poor neighborhood, Donna had the loving support of her family. Donna's parents divorced when she was a young girl. Her mum raised her and her older sister with the help of their grandparents. They would share their love, laughter, and joy of learning with Donna throughout her childhood.

Donna shared her loving home with others, as well. "There were kids on my street who had really difficult lives. I remember having them over to stay, and **trying to be a good friend to them** . . . I invited them over a lot, and we shared snacks and played, and explored. I remember putting on music to cheer them up and we would dance."[1]

When she was seven years old, Donna recalls being extra kind to one of her friends after his mum died. Later when she was older, and her mum died, she treasured all of those who helped her through that sad time.

When Donna was just three or four years old, her mum began attending university to become a virologist, a research scientist who studies viruses, such as the flu. Donna's mum was her "greatest role model,"[2] encouraging her curiosity and questioning nature. Donna felt like her mum had been a scientist, even before she went to school to become one. **"She just questioned everything in a polite, curious way,"** Donna said. "I grew up learning that from her . . . She taught me 'to not take things for granted but to question why things are the way they are and why things happen.'"[3]

Donna loved the magic in books, such as in the Harry Potter series and in *The Lion, the Witch and the Wardrobe*, but found real magic in how a microscope magnifies objects. Whether it was a leaf, a speck of dust, a dog hair, a crumb of bread, or a piece of lint placed on a slide, the microscope was an astonishing tool for discovery. Her mum, who used one every day in a laboratory, shared Donna's enthusiasm as they studied things together. Later on, microscopy would be an important part of Donna's studies, too, as she prepared to become a research scientist, like her mum—Dr. Joy Kean.

"This was one of the places where I developed a love of being in nature."[4]

KINGFISHER

Donna was curious and loved to explore nature. Her nana and papa, her dad's parents, lived in Netherlee, an area south of Glasgow, by the White Cart River. Donna often read and played inside her flat surrounded by buildings and drab streets. Netherlee, on the other hand, had green, lush woods where she felt free to run and play games with her sister, a sparkling river to throw stones into, with creatures both on the ground and in the sky. She was always on the lookout, on low-hanging branches by the river, for the kingfishers who would dive in the water for their dinners. Kingfishers, with their iridescent blue back and copper chest, were her nana's favorite bird.

Donna and her sister loved finding different creatures in between the rocks at low tide.

Donna's family also found more spectacular nature in Scotland seven hours away from Glasgow by car and ferry, travelling to the Isle of Harris on the Atlantic Ocean. Donna remembers jumping into the sand, creating sandcastles, and making dams as the tide came in. It brought with it bladder wrack, scallop shells, razor clams, and whelks, which lined the shore. Some of the Earth's oldest rocks are found on the Isle of Harris.

Around Glasgow, there was hiking and bicycling at local parks. More creatures could be found at one of her favorite places— Calderglen Country Park, which she loved to visit with Gran, her mum's mum. In its tropical conservatory, **she recalled feeling as if she were "in the jungle** all of a sudden."[5] There were lizards and birds, but she was most charmed by the leaf cutters marching back and forth along a rope near the ceiling, carrying enormous leaves in their jaws. Edinburgh Zoo was a rare treat as it was

LEAF CUTTER ANTS

expensive for her family. She was fascinated by its monkeys and apes. She found them fun and lively as she watched them jumping from one branch to another, making sounds.

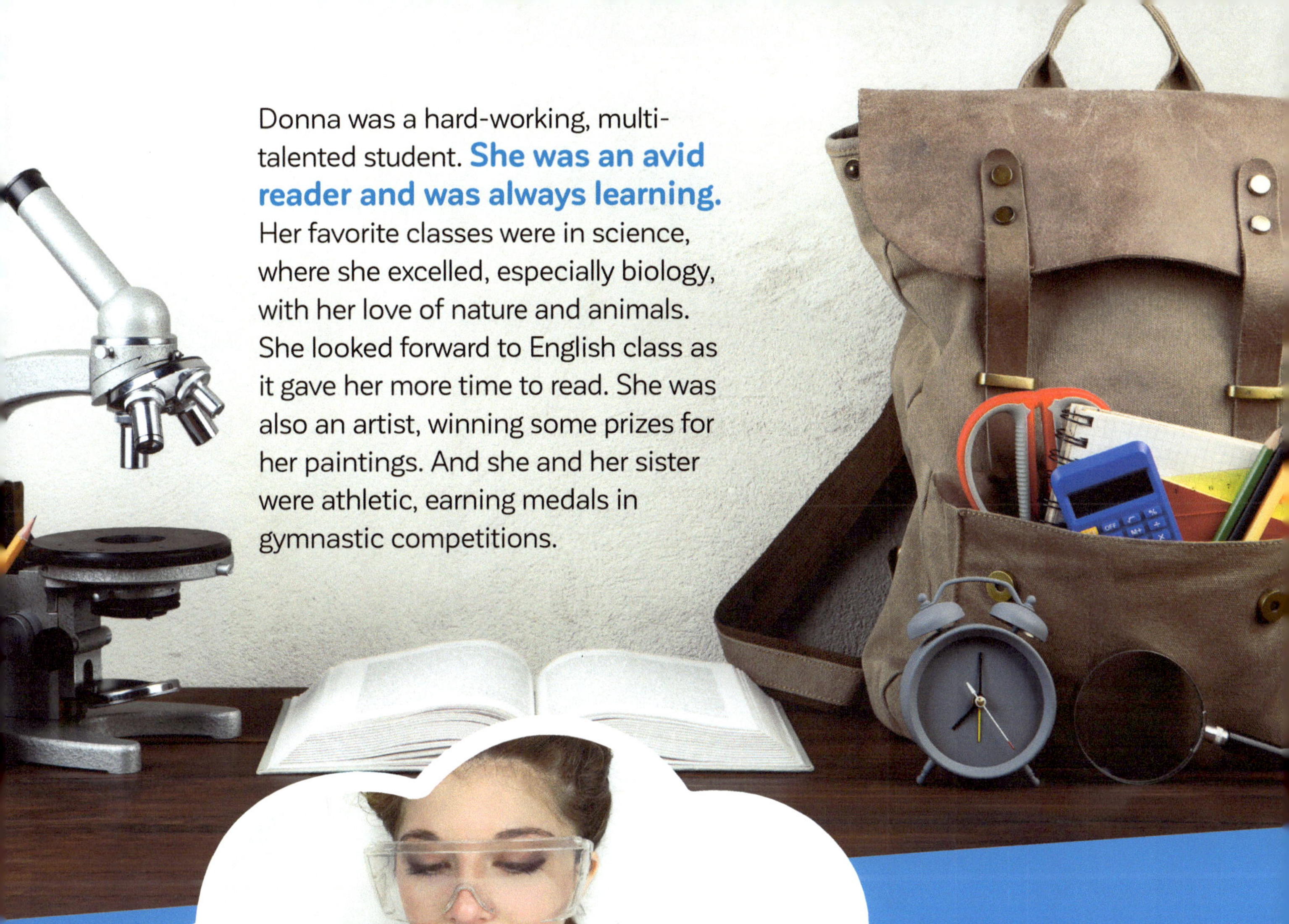

Donna was a hard-working, multi-talented student. **She was an avid reader and was always learning.** Her favorite classes were in science, where she excelled, especially biology, with her love of nature and animals. She looked forward to English class as it gave her more time to read. She was also an artist, winning some prizes for her paintings. And she and her sister were athletic, earning medals in gymnastic competitions.

After school, as a teenager, Donna would sometimes wait for her mum to finish work at her laboratory. Donna didn't mind the wait and found the lab to be an "amazing" environment. **It was fascinating to watch her mum and the other scientists** work on experiments. On occasion, they would let her wear a lab coat, which made it seem more real to her that someday she could become a scientist, too.

During her high school years, Donna gained self-confidence in public speaking. Her mum encouraged and supported her as she practiced her talks for school assignments. Donna remembers particularly when her mum helped her when she had to share a report on a two-week work experience at her former elementary school, Thornliebank. Donna earned the top public speaking grade in her class for her presentation. Before she graduated from high school, Donna was honored to be chosen, along with one other student, to present ideas in front of all of her peers from the local high schools.

Donna had once thought that she would become a schoolteacher as she loved sharing ideas and working with children. **But the role of a scientist was beckoning her.** Later, she realized she was a teacher, sharing the results of her experiments, and loving it.

Donna was **fortunate to be able to pursue her passion for animals** in her university studies. Since she was born in Scotland, she could follow her dreams without significant financial worries. Her first four years of university were free. Her aunt helped her pay for her later years in school. At the University of Strathclyde, in Glasgow, she majored in psychology, studying communication, first in children and then in chimpanzees.

During her third and fourth years at the University, Donna gained invaluable experience and boosted her skill levels and self-confidence in her dream to become a scientist. She volunteered to be a lab assistant during the school year. Enthusiastic and trustworthy, Donna diligently ran hours and hours of experiments for a research project, until they felt second nature to her. The following year, due to her experience and dedication to her work, she was hired as a research assistant. As the president of the Psychology Society, she led the other students in sharing new research and learning opportunities.

She was intrigued by Dr. Jane Goodall's study of chimpanzees in the wild, in Tanzania, and how they used calls and gestures to communicate with each other.

DR. JANE GOODALL

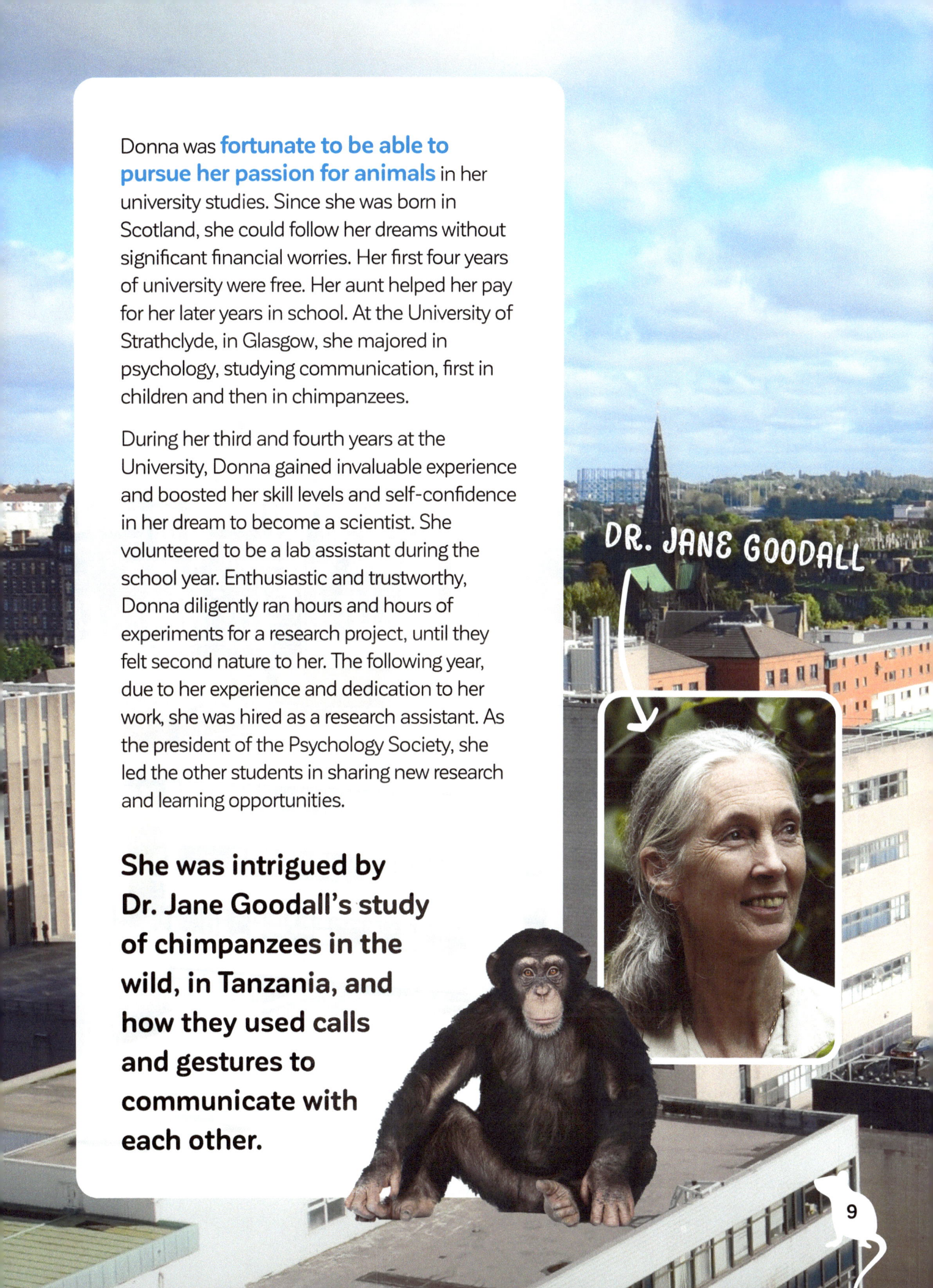

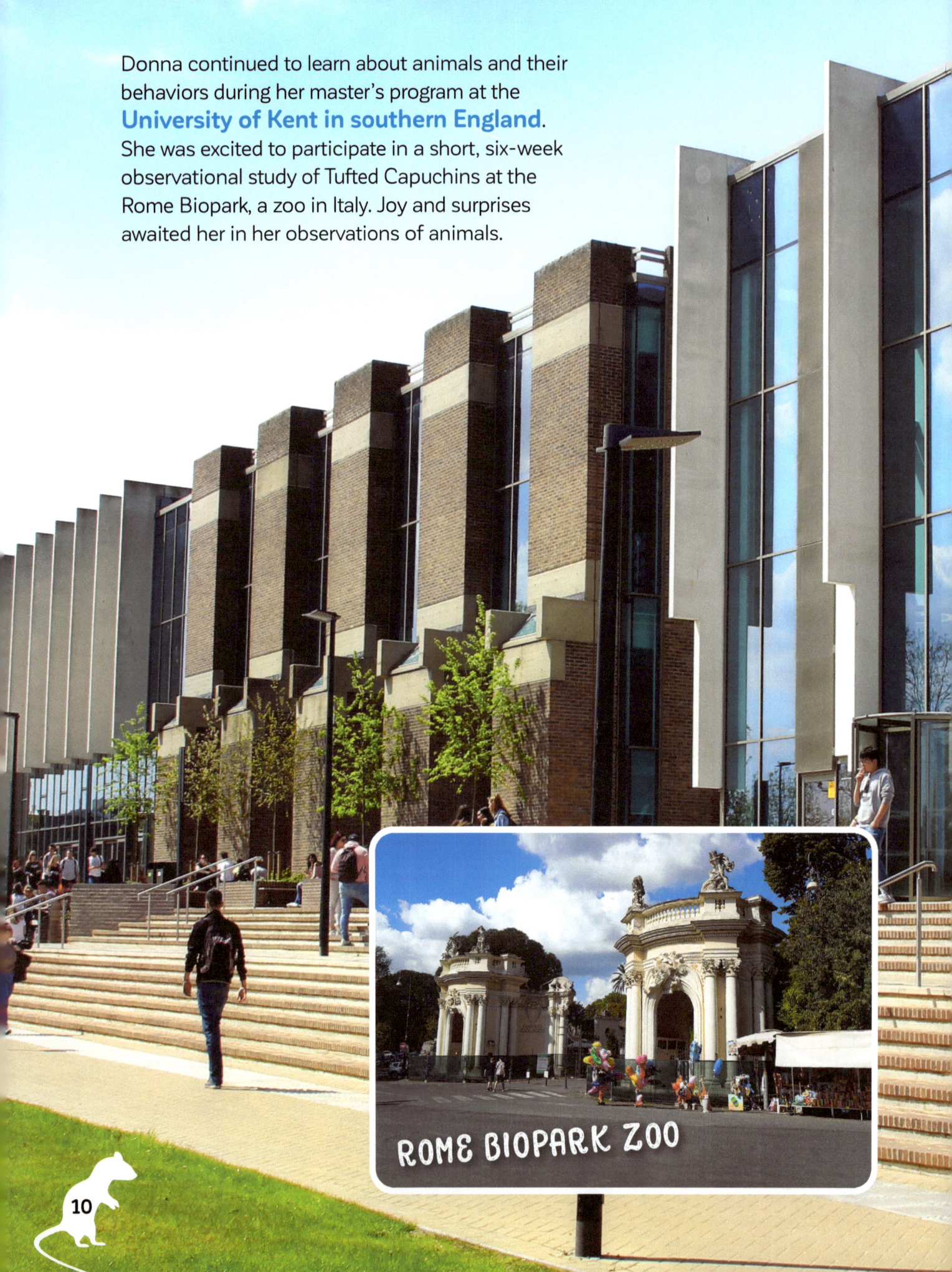

Donna continued to learn about animals and their behaviors during her master's program at the **University of Kent in southern England**. She was excited to participate in a short, six-week observational study of Tufted Capuchins at the Rome Biopark, a zoo in Italy. Joy and surprises awaited her in her observations of animals.

ROME BIOPARK ZOO

Donna and other researchers tested the hypothesis that the capuchins would scratch themselves more when they made an alarm call—a call usually made if there was a predator nearby. When capuchins are anxious, they scratch themselves. Perhaps you know someone who taps their foot or twirls their pencil when they are anxious.

Donna **observed the capuchins and collected data** on a chart about their behavior. She wrote down the date, time, number of alarm calls, and number of times the capuchins were self-scratching.

The sample data collection chart uses the 24-hour clock.

Date	Time	# of Alarm Calls	# of Self-Scratching
4/11/2014	9:15	1	3
4/11/2014	10:35	1	4
4/11/2014	13:07	1	2

Monkeys are unpredictable!

One day, Donna moved too close to Carlotta, a capuchin monkey. All of a sudden, Carlotta grabbed one of Donna's papers. There was a tug-of-war! The paper was torn in half, and Carlotta quickly gobbled her half. After her initial shock, Donna laughed and was reminded how clever these monkeys are. And, she had learned not to sit too close to them. You never knew what could happen.

Donna analyzed her data and shared it with her colleagues, the other researchers working on the same project. The hypothesis was proven to be true. Capuchin monkeys scratched themselves more when they were making the alarm calls. **Donna co-wrote her first professional publication** based on this observational study. It was a joy to have finally taken part in a real-life study of animal behavior.

CAPUCHIN

Donna persisted in her studies at the University of Stirling, in central Scotland, to become a behavioral research scientist. She would earn her PhD from her work at this university, earning her the title of "Dr." before her name. For her research project, she would study how young human children, monkeys, and apes were able to carry over new learning from one task to another on a touchscreen computer. It was a new, **innovative research project**—using touchscreen computers with monkeys and apes.

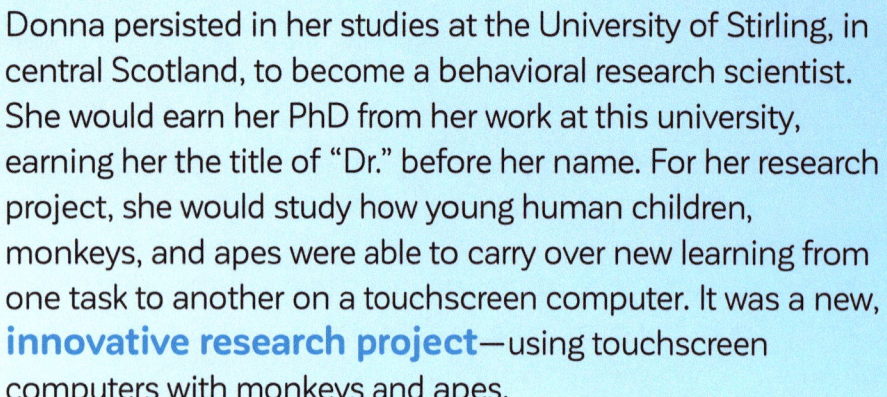

↑
CHIMPANZEE

On an island near the University of Stirling, Donna assembled **her first, small research station** for experiments with chimpanzees and macaques. She found it an incredible experience. She had only three months to experiment with the chimpanzees, while they were in an indoor enclosure. She learned that it wasn't enough time to train them. And, on occasion, they would surprise her. "With absolutely no warning," they threw poo at her.[6]

MACAQUES

Donna was able to work eight to nine months with the macaques. Some of them were trainable on the touchscreen computer, but unfortunately the project ended early. She was disappointed to not have the chance to finish her experiments.

Donna experimented with Edinburgh Zoo Tufted Capuchins at the beginning and at the end of her research project, working at the Edinburgh Zoo, Living Links Center. Although scientists had experimented with capuchins before, this was the first time capuchins had interacted with touchscreen computers. She found that the **capuchins were easy to train** on the computers compared to the other monkeys and apes she worked with. As the capuchins worked on tasks, they were given rewards for correct choices and no reward for incorrect ones.

The capuchin is reaching out to touch the computer screen.

Sometimes Donna stayed overnight at the zoo, often waking up at dawn to the sound of "gibbon song."[7]

All scientists doing research at Living Links are trained in how to respect the monkeys' choice of participating or not in research. When the capuchins chose to be part of Donna's study, they went into a cubicle from their indoor or outdoor enclosure. When they wanted to leave the study, they would give a signal to Donna and be let out of the cubicle.

From Donna's blog, *"A day in the life of my PhD: when your participants are monkeys."*

"…One of these groups (of monkeys) included a band of particularly rambunctious young capuchins who lived to run amok during these research sessions. They would fly in and out of the research cubicles, sometimes luring me into believing they wanted to participate only to dart out at the last moment.

To separate, a monkey would enter the cubicle and allow me to close sliding doors that divided them from the tunnels linking the cubicles to their enclosures. Often, the nonsense crew would not take part at all but would loiter around the cubicles and insert their tiny fingers between the partitions and the wall, so that the doors couldn't be closed and thus no-one else could take part either. Sometimes they would put on a wrestling display; sometimes they would play what seemed to be Tag, racing through the cubicles.

When a monkey had allowed me to separate them from the group and was ready to engage in the task, I would check which stage of the experiment they were on… Just like humans, there is a lot of variation in individual monkeys' behaviour and abilities, and the most enthusiastic players are not always the best performers (but are usually my favourites).

After cleaning the monkey prints from the screen and tidying my apparatus away, I would have lunch with other researchers and zoo staff and update my records. Then repeat for the afternoon session…"[8]

DONNA'S MUM

During Donna's PhD program, her mum was diagnosed with cancer. It was a challenging time for Donna. She worried about her mum and tried to stay focused on her research. Fortunately, the cancer medications seemed to work for about two years. **During this time, both Donna and her sister planned their weddings,** so their mum could celebrate with them.

When the cancer treatments were no longer working, Donna left school and took care of her mum. **Her beloved mum passed away** three years after the beginning of Donna's PhD program. Donna's heart was saddened. She took time off from school to heal.

Time passed. Donna healed with the help of her loving husband, friends, hikes in nature, music, and calming yoga movements. She would always miss her mum and treasure the memories of her. **"There will be good things again,"** she thought hopefully.[9] She would become a research scientist like her mum, and her sister was going to have twins.

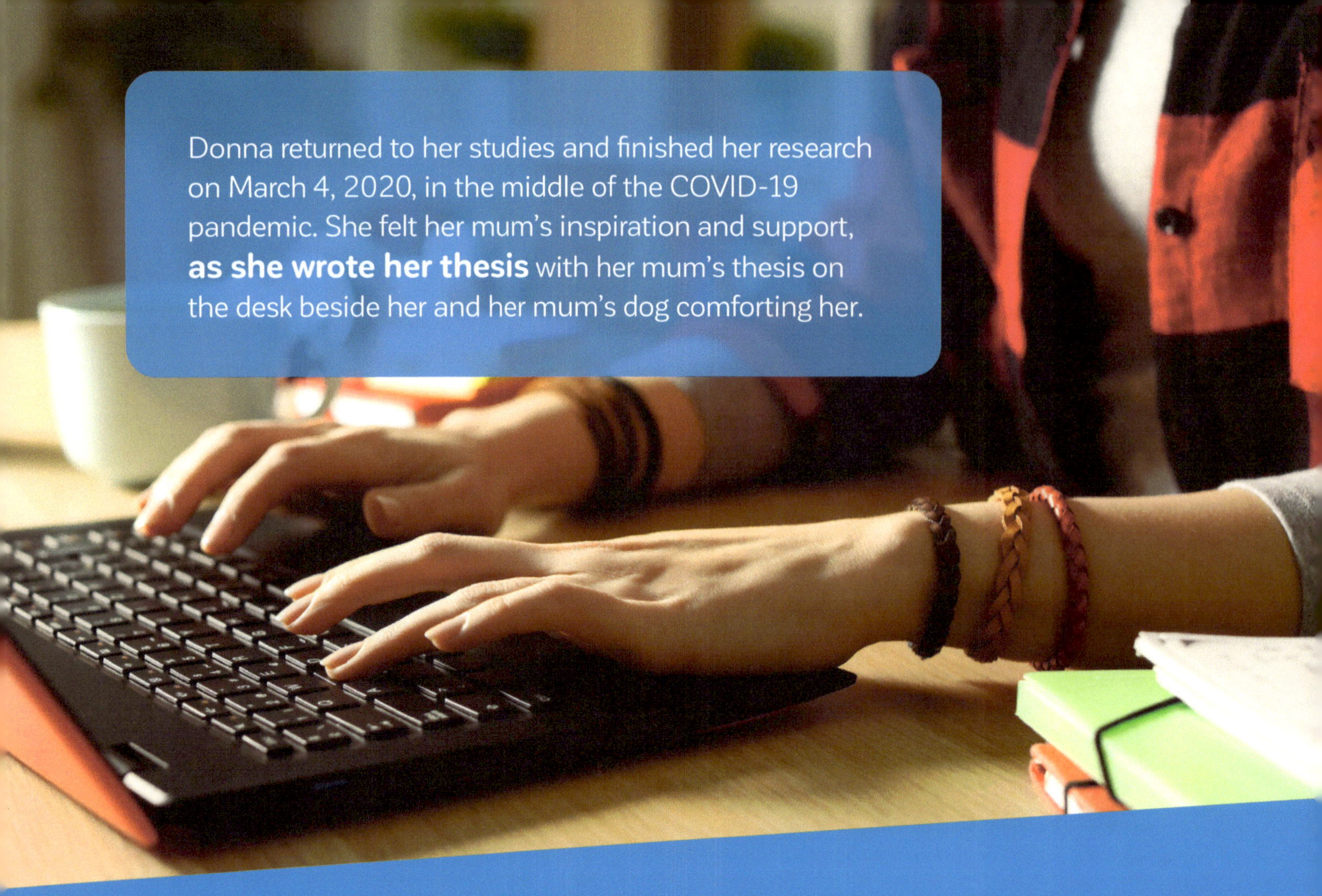

Donna returned to her studies and finished her research on March 4, 2020, in the middle of the COVID-19 pandemic. She felt her mum's inspiration and support, **as she wrote her thesis** with her mum's thesis on the desk beside her and her mum's dog comforting her.

Donna recorded her first podcast and presented her research in North America and Europe.

Results from Donna's and her colleagues' experiments showed that children had the most success with carrying over their learning on the touchscreen computer tasks. The capuchins performed the best of all of the animals studied. The study resulted in five research articles, a paper at a conference, and a poster.

Donna graduated in March 2021 from the University of Stirling, earning her PhD. She was following in her mum's footsteps. Her mum's love and scientific way of looking at things would always be with her. Donna searched and was fortunate to find an ideal behavioral research position, one which uniquely blended her love of science and animals.

Donna began a new adventure in April 2021.

She was hired by APOPO for exciting and groundbreaking training of African giant pouched rats. APOPO is the acronym for a Belgian non-profit organization, which has found a way for humans to benefit from rats' and dogs' keen sense of smell. APOPO's goal is researching and developing scent detection applications for these animals to save human lives and improve the environment.

SAVES LIVES

When Donna first saw the advertisement for a position at APOPO, she wasn't sure if it was real—the work sounded so unusual. Her phone call to them changed her life. She shared with her family how this position was such a good match for her. In an interview with APOPO for The International Day of Women and Girls in Science (2022), Donna shared, "I wanted to do research that translated into real-world benefits for people—the humanitarian work we do here at APOPO certainly ticks that box!"[10]

APOPO has its main offices and rat training centers in Morogoro, Tanzania, on the campus of the Sokoine University of Agriculture. These centers include the landmine and the tuberculosis centers, as well as the new search and rescue training site, at which Donna would be working. **Donna and her husband moved from Scotland to Tanzania for her new work.** They took two flights, traveling from Scotland to Dar es Salaam, Tanzania. Then they journeyed by car from Dar es Salaam to Morogoro, passing by small towns, fruit and cashew trees, and long stretches of Miombo woodland.

SCOTLAND

TANZANIA

CASHEW FRUIT

They settled into their new home in Morogoro. It was the first house Donna had ever lived in. She had only lived in flats before. **"It's been amazing to come here and have this lovely big house and giant garden,"** Donna shared. "We have this tropical garden, and there are all these mature fruit trees and beautiful plants . . ."[11] In the garden, she would grow vegetables (carrots, spring onions, and bell peppers), herbs (coriander, basil, and lemongrass), and even sugarcane, which she learned was an important crop in Tanzania.

SUGAR CANE

And, she was surprised to see kingfishers again, although the ones in Tanzania are not the same species as the ones in Scotland. Memories of nana's beloved bird helped her feel more at home in this new country.

Although, Donna was excited to jump right into her new position, **she first needed to read and study APOPO's history and procedures**. She needed to learn how to work best with the African giant pouched rats and how they are trained to earn their title of HeroRAT, an important title given only to those who finish their training and are tested and certified for a high level of accuracy in their work. She was so looking forward to being a part of their new training.

Donna learned that the idea for APOPO started with a Belgian man named Bart Weetjens. As a child, he raised pet rats and other rodents. He had learned first-hand about their keen sense of smell and how intelligent they were. As an adult, Bart read about how gerbils had been used as scent detectors for landmines. **He wondered if rats could help with the problem of landmines in post-war countries.** In 1995, he began to explore this idea. He met with Prof. Ron Verhagen, an expert on rodents, in March 1997. African giant pouched rats were recommended for landmine detection. They were common in Africa, inexpensive to raise, and could live up to eight years. On November 1, 1997, Bart and a former classmate, Christophe Cox, started APOPO in Morogoro, Tanzania.

BART WEETJENS

APOPO decided to raise the African giant pouched rats, which it would use in its training. African giant pouched rats and pet rats are from different rodent families. The African giant pouched rats are similar to hamsters and gerbils with their pouched cheeks. Pet rats live for about two to three years and their size can vary. Some can grow to be about 45.72 centimeters (one and a half feet in length) including their tails. African Giant Pouched Rats, on the other hand, live about eight years in captivity, and grow to 60.96 to 91.44 centimeters (two to three feet long) including their tails.

Well cared for—the rats would be asked to engage in important work. They would work only in the early mornings, as they are sensitive to sunlight. They would receive a healthy diet, wellness checks, and have time for regular playtime where they can dig, climb, and explore. When the rats are retired from their work at APOPO, there is a special place created for them, called a sanctuary. They are treated with care and respect and able to comfortably live out the rest of their lives.

When the baby rats are four to five weeks old, they begin what is called socialization. **They need to become used to being around humans.** APOPO workers, called handlers, are carefully instructed how to best train the rats. The handlers are gentle with the rats, teaching them about textures, such as the difference between crawling in their cages and on the ground. The handlers also help the rats to acclimate to sounds around them without being startled, such as cars beeping and motors running. The sight and smell of different people becomes second nature to the rats, too. And most importantly, they learn to trust the handlers, that they will be kind and gentle with them. As a young child gradually becomes used to their home, so the rats also adjust to their APOPO home.

SAMIA & SAID

NEEMA & SHURI

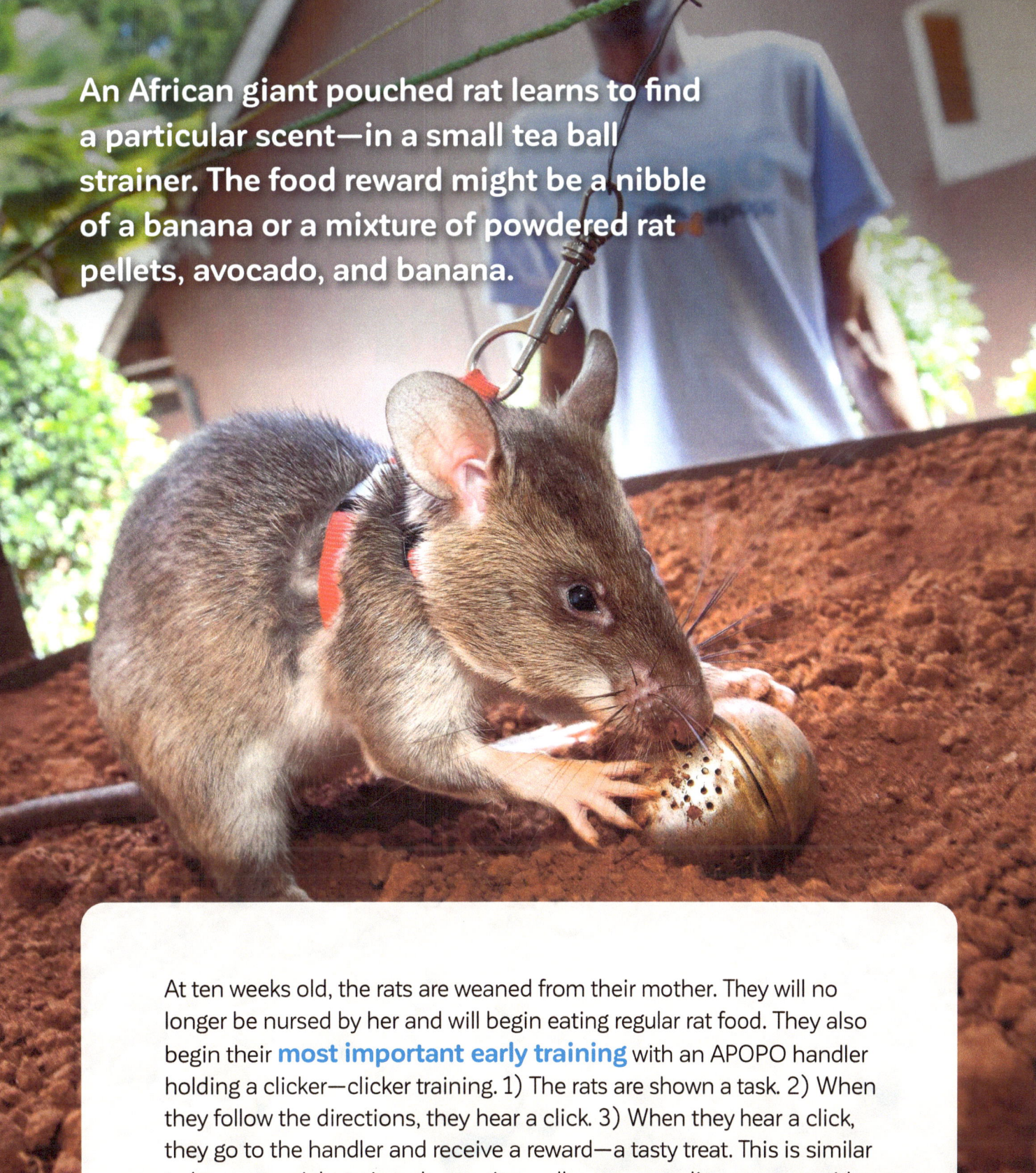

An African giant pouched rat learns to find a particular scent—in a small tea ball strainer. The food reward might be a nibble of a banana or a mixture of powdered rat pellets, avocado, and banana.

At ten weeks old, the rats are weaned from their mother. They will no longer be nursed by her and will begin eating regular rat food. They also begin their **most important early training** with an APOPO handler holding a clicker—clicker training. 1) The rats are shown a task. 2) When they follow the directions, they hear a click. 3) When they hear a click, they go to the handler and receive a reward—a tasty treat. This is similar to how you might train a dog to sit or rollover—rewarding your pet with a food treat after the action is done. Once the rats learn the connection between hearing the click and receiving a yummy treat, they are ready to participate in more specialized training.

Scent detection training begins with the rats learning to recognize a particular scent. Donna learned how the African giant pouched rats at APOPO were **first trained for detecting the scent of TNT**—the smell of explosives used in landmines. Explosives are dangerous. They are hidden in the soil and can injure people and make land unusable for living. People can't grow crops and children can't play safely in areas that might have landmines.

The African giant pouched rats take nine months to a year to complete their training on the training minefield. It was created in 2002 by APOPO with support from the Tanzanian Army in Morogoro.

Wearing a harness around their bodies, the rats are connected to a pulley system between two APOPO handlers. The rats have learned to go back and forth over a specific area of land. When they detect the smell of a landmine, they stop. They scratch at the ground. If they have correctly detected a landmine, they hear the "click" sound from a handler. They return to the handler and receive a treat. The landmine spot is marked. Later, the landmine is removed and exploded in a safe area.

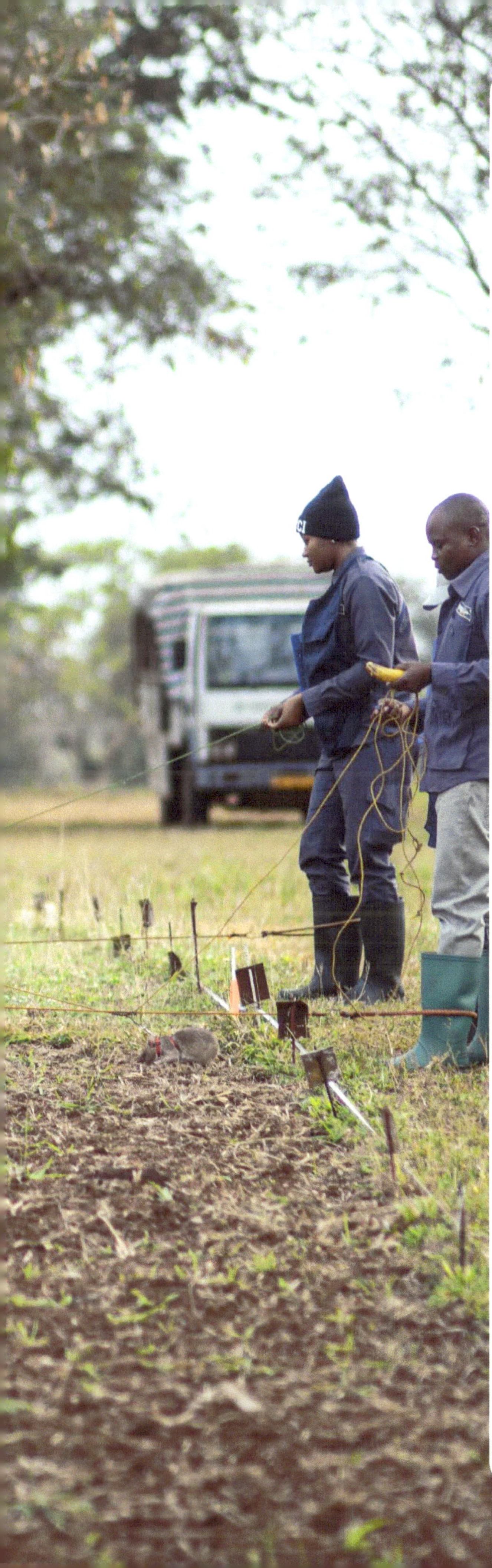

After repeated testing and certification for accuracy in detecting TNT, **the rat is called a HeroRAT**. Between 1997 and 2015, the HeroRATs helped to clear 13,000 landmines in Angola, Cambodia, and Mozambique. By ridding areas of landmines, APOPO restores land once considered dangerous to a safe and productive use. The land can now be used for building homes, growing crops, providing a pathway to water or a safe place for children to play. As of November 1, 2022, APOPO's 25th anniversary, APOPO had cleared 17 million square meters of land and destroyed over 150,000 explosives.

Working quickly, the HeroRATs can check a rectangular area measuring approximately 23.47 meters (seventy-seven feet) long and 7.92 meters (twenty-six feet) wide, in thirty minutes. It could take a human trained to remove explosive mines up to four days to do the same work.

Next Donna studied how the HeroRATs were trained to detect tuberculosis (TB), beginning in 2007. Tuberculosis is the second highest infectious disease taking lives all over the world, after COVID-19. It is caused by a special type of bacteria that destroys lung tissue. When someone has TB, they can easily spread it by coughing, singing, or sneezing.

APOPO has partnered with local clinics, working to provide what is called second-line screening, using trained HeroRATs. People in sub-Saharan Africa, who have symptoms of TB, go to a medical clinic and are usually asked to give two sputum samples. This is done by taking a deep breath and coughing hard until they can spit fluid from their lungs into a collection cup. The sample is put on a slide to be checked twice, once using a microscope at the clinic, and a second time by being "tested" by APOPO HeroRATs. The samples include both TB-positive and TB-negative results. Upon arrival at APOPO, all of the samples are treated with heat so the APOPO trainers and HeroRATs cannot be infected with TB.

Using their powerful sense of smell, the HeroRATs have learned how to recognize the smell of tuberculosis in samples of sputum. The samples, which the clinics found to be TB-positive, are used as control samples to teach the rats. When the rats smell TB in the samples, they let the trainer know by pausing and sniffing for about three seconds over the sample. When they are accurate with detecting TB, they hear a click and receive a treat.

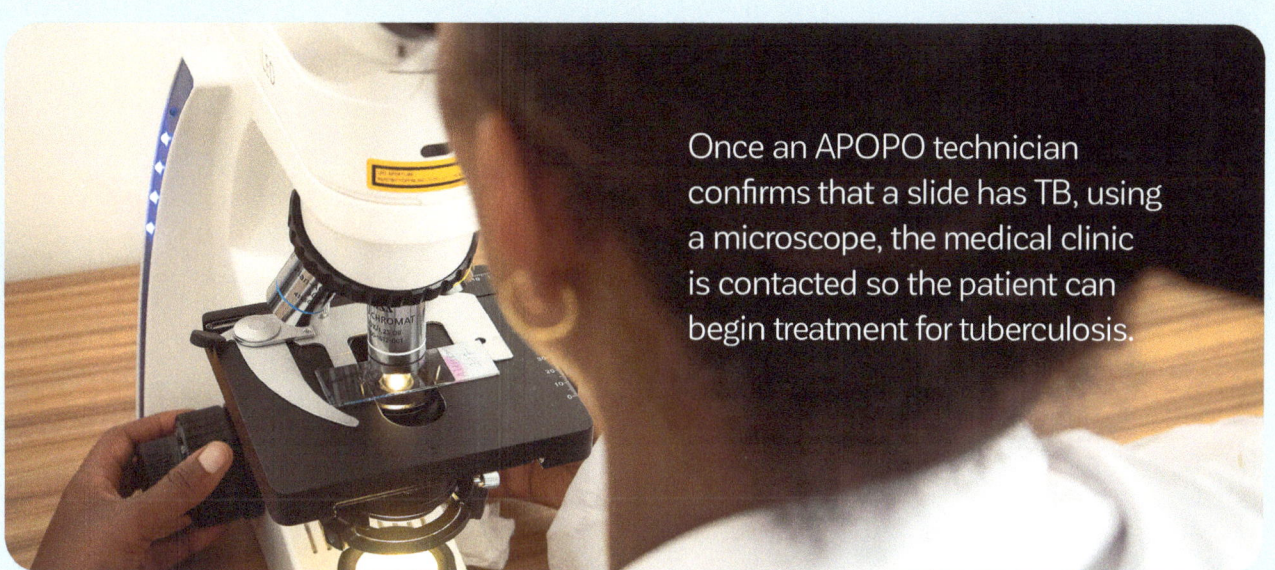

Once an APOPO technician confirms that a slide has TB, using a microscope, the medical clinic is contacted so the patient can begin treatment for tuberculosis.

The HeroRATs work quickly and can screen one hundred sputum samples in under twenty minutes. It can take a lab technician all day to review just twenty-five sputum samples using a microscope.

HeroRATs have the potential to increase the accuracy of TB detection in the clinics by forty percent. They are finding patients with TB, who were missed by clinics. As of APOPO's 25th anniversary, HeroRATs have screened 850,000 slides of sputum samples, detected 150,000 cases of TB, and stopped 200,000 possible TB infections.

After studying what APOPO had worked on so far, Donna was ready to begin training these marvelous rats. From the very beginning, Donna was comfortable with them and surprised by how curious, intelligent, clean, and gentle they were. "They're quite misunderstood," she said. "They're lovely animals and very friendly. It's all about how you treat them."[12] She remembered how her dog Zack had been wild at first, yet when trained by her family, he had become such a loving dog.

Donna was excited to contribute to APOPO's Innovation Team by leading the search and rescue training of the African giant pouched rats. **They would be known as RescueRATs once they finished their training.** But others loved to call them "rats with backpacks." Donna would be figuring out the steps to train them in a new way to save people's lives. This was the most challenging work for the APOPO rats so far. Donna loved having this opportunity!

The idea for APOPO rescue rats came about when Turkey's GEA search and rescue team contacted APOPO to see if their rats could be trained to search collapsed buildings for survivors after natural disasters. **The country of Turkey is located on two fault lines and is prone to earthquakes.** Dogs were already part of Turkey's search and rescue team, but perhaps APOPO's rats could aid their research efforts, due to their keen sense of smell and smaller size. The rats would be able to access areas that dogs were too big to go into.

The search and rescue project "is the most complicated project APOPO has worked on with the rats so far," Donna said.[13] Nine African giant pouched rats were chosen for this project. Only seven were able to complete all of the challenging training steps—Kiria, Jo, Pearce, Caruso, Billy, Daniel (named after Dr. Donna Kean's nephew), and Wagner (named after a researcher in animal training research). APOPO built a mock debris site in August 2021, for training the RescueRATs. **The building was designed to mimic a debris site following an earthquake.**

DANIEL

Similar to the rats trained for landmine and TB detection, the search and rescue rats are bred at APOPO. They begin their socialization at four weeks old and their training at four months old. Unique to their being trained to become RescueRATs, they will begin to wear a vest as a baby. As they grow, they'll wear bigger vests to be eventually replaced with a tiny backpack.

31

When the rats were four months old, they began their search and rescue training at the debris site. It was empty at the beginning of the training.

STEP 1 The handler releases a rat from a place called a basepoint, an opening on the side of the debris site. This is a spot where the handlers stay during the training. The rats go into the debris site and wander around. When they hear a beeping signal from the handler, they return to the basepoint and receive a treat.

STEP 2 The rats leave the basepoint and, using their sense of smell, detect a human acting as a "survivor" in the debris site. Once the survivor is found, the rats learn to pull on a ball, attached to the vest at their neck, then they receive their treat from the person acting as the survivor. The ball is a cleverly designed microswitch with a trigger. The rats will not have to return to basepoint for this step.

STEP 3 Steps 1 and 2 are put together. The rats leave the basepoint, learn to pull the ball when they're near a person, and then return to basepoint and receive a treat.

Gradually the rats' debris site will become even more complex. The search base enlarges, and more rooms and debris are added. A second story is added to the debris site to mimic a damaged multi-level building. Rescue operation sounds are also added, such as drilling or digging, so the rats become familiar with a more realistic post-earthquake setting.

The people acting as survivors also further challenge the RescueRATs in training. The "survivors" are asked to not move about or make any noise so the rats can only use their sense of smell to find them. The "survivors" also hide themselves so they are trickier to find, such as underneath or behind furniture. Then new people are added as survivors. These new people are a challenge for the rats, because they haven't learned their scent yet. In the future, the rats will be trained to find survivors in new environments away from the debris site.

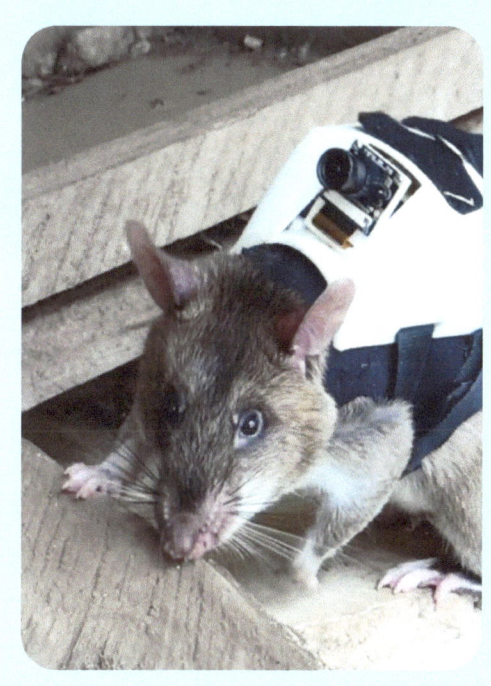

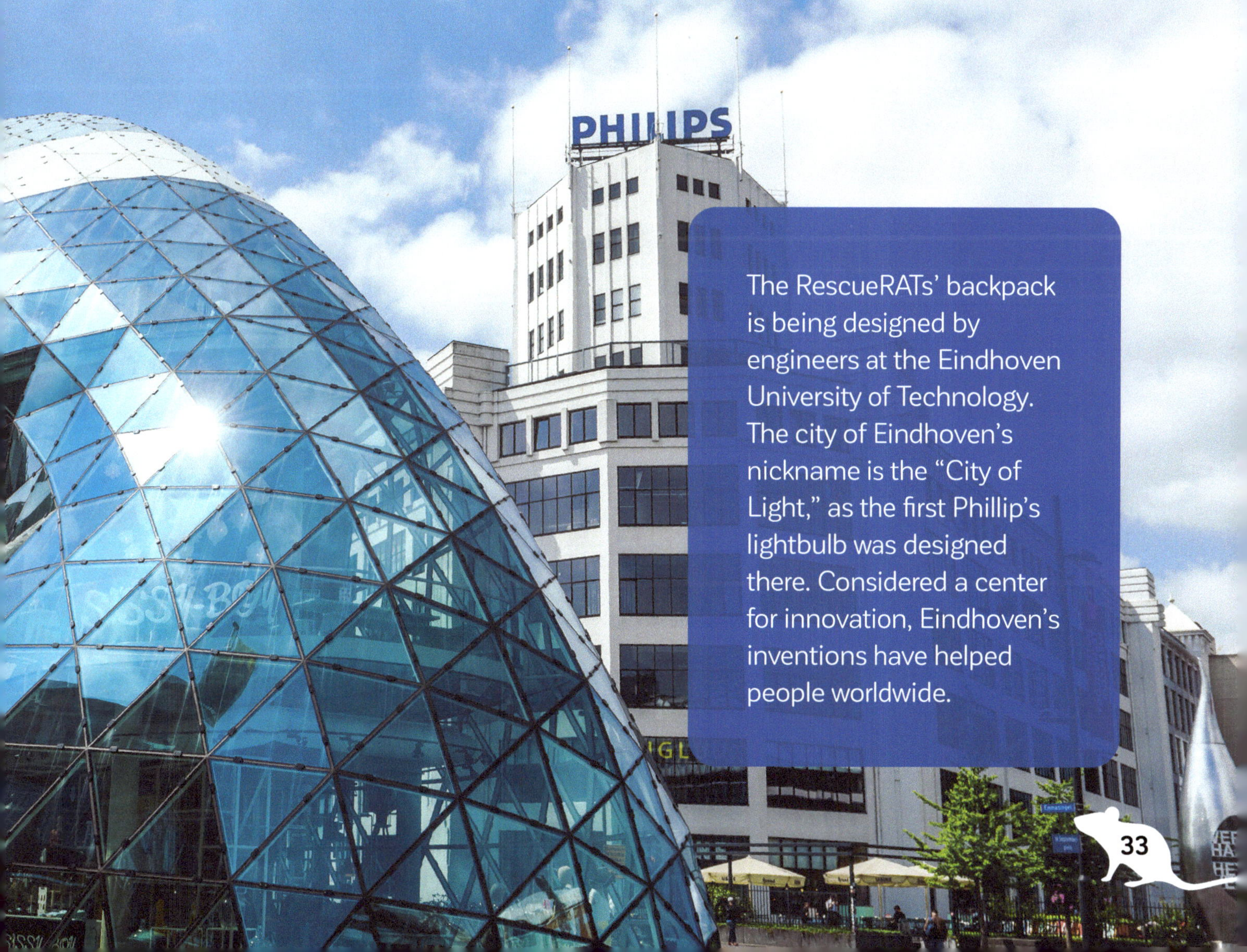

The RescueRATs' backpack is being designed by engineers at the Eindhoven University of Technology. The city of Eindhoven's nickname is the "City of Light," as the first Phillip's lightbulb was designed there. Considered a center for innovation, Eindhoven's inventions have helped people worldwide.

The RescueRATs have begun training with their second backpack prototype. **Engineers continue to work on the design of the backpack** and its specialized technology. It will have four important features. 1) A video camera so rescuers can see the survivor and if they have any injuries. 2) A GPS-type system for finding the location of the survivor for rescue efforts. 3) A two-way radio for communication between the rescue workers and the survivor. 4) A greeting from the rescue workers to comfort the survivor when they see a rat approaching them.

Once the rats complete their search and rescue training, they will be called RescueRATs. When they are ready to be deployed to Turkey, they will be accompanied by one or two people from APOPO, perhaps a scientist and a research assistant, who will work closely with their search and rescue partners in Turkey. Based on the experience during deployment, the APOPO trainers will change the search and rescue training, so the RescueRATs are even better prepared for their next chance to help with the search and rescue operations.

Leading the APOPO Innovation Team was a remarkable experience for Donna. "I would never have believed that I could do the job I'm doing now, coming from where I came from."[14] As a girl, she couldn't have imagined leading the APOPO search and rescue project. Although she grew up in a poor neighborhood, Donna had the loving support of her family, the opportunities to nurture curiosity, a love of learning, and her mum as a role model for becoming a research scientist. Donna had steadily built up her confidence and her belief in herself from childhood through her university years, so she could contribute to the remarkable, humanitarian work at APOPO.

There is much growth ahead for Donna and for APOPO.
APOPO has made a significant difference in people's lives in its first twenty-five years. It will continue to provide training to HeroRATs for landmines and tuberculosis detection. Its innovative work in the newer scent detection area of search and rescue will save more lives in the future. Who knows what their next project will be.

More about Dr. Donna Kean

On February 11, 2022, Dr. Donna Kean was interviewed by APOPO in honor of the International Day of Women and Girls in Science. She shared why she became a scientist. "I chose science because it is such a varied and rewarding field. Not only do you get to read and write about interesting, cutting-edge topics, but you also do very practical work when conducting experiments. On top of this, there is problem-solving, grant writing, presenting at and travelling to exciting places for conferences, and meeting lots of like-minded people. For someone like me that is curious and likes variation at work, science is a great choice."[15] She encourages girls to pursue their love of science. "Believe in yourself and the value you bring to the table. As long as you work hard and do your research, nothing can stop you." [16]

For more information about Dr. Kean's publications (articles, poster and conference paper) go to: https://www.researchgate.net/profile/Donna-Kean

Dr. Kean is now experimenting with parrots and rats in a postdoctoral position at the Autonomous University of Barcelona's (UAB) Institute of Neuroscience. For more information about her current research project at the UAB Animals Mind Lab go to: Testing for the universal mind using probabilistic inference—Universitat Autònoma de Barcelona Research Portal.

More About APOPO

APOPO is a Dutch acronym for Anti-Persoonsmijnen Ontmijnende Product Ontwikkeling. This translates to Anti-Personnel Landmines Product Development. APOPO's website (https://apopo.org) is an invaluable resource for learning about APOPO.

APOPO celebrated its 25th birthday on November 1, 2022. A global non-profit and Belgian NGO (non-governmental organization), it is dedicated to saving lives through its humanitarian work. After highly rigorous training, the African giant pouched rats (nicknamed HeroRATs) are certified in detecting landmines and other explosive, as well as diseases such as tuberculosis. Its research and development department is exploring other applications for its HeroRATs. This book shares APOPO's developments in training rats for Search and Rescue.

APOPO is also training rats to detect smuggled animals (pangolins, for example) and other wildlife products. In July 2024, Dr. Jane Goodall, an APOPO Friend and Advisory Board member, visited APOPO and named a Wildlife Detection HeroRAT in training "Jane." There are currently ninety-six rats trained to detect landmines. Thirty-two rats are trained to detect tuberculosis. Forty-four rats are working with research and innovation.

APOPO also has a HeroTREEs project, which focuses on the environment. For further information about this project, see www.apopo.org/herotrees.

There is an APOPO Visitor Center in Siem Reap, Cambodia where African giant pouched rats can be observed detecting TNT at a "mock" landmine area. African giant pouched rats can be observed in scent-detection demonstration zookeeper talks in the United States at the Point Defiance Zoo and Aquarium in Tacoma, Washington; at the Indianapolis Zoo in Indianapolis, Indiana; at the Maryland Zoo in Baltimore, Maryland; and at the San Diego Zoo in San Diego, California.

Dr. Kean is now experimenting with parrots and rats in a postdoctoral position at the Autonomous University of Barcelona's Institute of Neuroscience. Dr. Cindy Fast is working as the Head of Training at APOPO. Expect to hear more exciting news about Dr. Fast's work with the RescueRATs as well as other amazing APOPO research projects.

A portion of this book's sales will be donated to APOPO. If you would like to contribute to its important humanitarian work, see its website: https://apopo.org/support-us/donate.

Endnotes

1. Donna Kean, SKYPE interview with the author, August 19, 2022; Donna Kean, email correspondence with the author, September 6, 2022.

2. APOPO, International Day of Women and Girls in Science, 2022, https://apopo.org/in-focus-dr-donna-kean.

3. Donna Kean, SKYPE interview with the author, August 19, 2022.

4. Donna Kean, SKYPE interview with the author, August 19, 2022.

5. Donna Kean, SKYPE interview with the author, August 19, 2022.

6. APOPO, International Day of Women and Girls in Science, 2022, https://apopo.org/in-focus-dr-donna-kean.

7. Donna Kean, "Some poignant moments during a PhD and my mum's battle with cancer," PhD Women Scotland, https://phdwomenscot.wordpress.com/2021/12/16/some-poignant-moments-during-a-phd-and-my-mums-battle-with-cancer/, 2021.

8. Donna Kean, "A day in the life of my PhD: when your participants are monkeys," PhD Women Scotland, https://phdwomenscot.wordpress.com/2020/04/16/a-day-in-the-life-of-my-phd-when-your-participants-are-monkeys/, 2020.

9. Donna Kean, SKYPE interview with the author, August 19, 2022.

10. APOPO, International Day of Women and Girls in Science, 2022, https://apopo.org/in-focus-dr-donna-kean.

11. Donna Kean, SKYPE interview with the author, August 19, 2022.

12. Donna Kean, SKYPE interview with the author, August 19, 2022.

13. Donna Kean, SKYPE interview with the author, August 19, 2022.

14. Donna Kean, SKYPE interview with the author, August 19, 2022.

15. APOPO, International Day of Women and Girls in Science, 2022, https://apopo.org/in-focus-dr-donna-kean.

16. APOPO, International Day of Women and Girls in Science, 2022, https://apopo.org/in-focus-dr-donna-kean.

Glossary

24-hour clock: A type of time telling where the hours are seen as "00" for midnight and "01" for 1:00 a.m., continuing the count up to 24 at the end of the day. 1:00 p.m., for example, would be 13:00.

anxious: To feel nervous, scared, or uncomfortable about something.

APOPO: A Dutch acronym for Anti-Persoonsmijnen Ontmijnende Product Ontwikkeling. In English, it translates to Anti-Personnel Landmines Detection Product Development. A global non-profit and Belgian NGO (non-governmental organization), it researches and develops scent detection applications for African giant pouched rats and dogs, to save human lives and improve the environment.

APOPO handler: A person at APOPO trained to work with the African giant pouched rats.

captivity: To be restricted to a certain area, such as animals in a zoo.

colleagues: Someone who works with you in the same type of work.

data: Specific information gathered, for example, in an experiment or observation, to answer a question in a study.

environment: Surroundings of a particular place or location, that could be provided by people or by nature.

European format for dates: The date begins with the day of the month first, followed by the month and year.

humanitarian work: Actions for helping other humans improve their lives, their health, and/or their happiness.

hypothesis: A question a scientist asks before beginning an experiment. Then information is gathered to prove the hypothesis either right or wrong.

master's degree program: More specific area of study than in an undergraduate program, for generally two years.

microscopy: The use of microscopes to study objects and areas around objects.

mimic: To copy something. In this case, the debris site at APOPO looked like or mimicked a building impacted by an earthquake.

oral syringe: A plastic syringe without a needle used to feed animals.

PhD: A one-to-two-year program in a specialized area of study. Upon completion of their study, the graduate is considered an expert in this specific field and will be addressed as "Dr." even though they may not have completed medical studies.

procedures: Steps involved in managing a project or task.

second-line screening: Referring to the tuberculosis screening, the tuberculosis slides are checked twice for accuracy.

sputum samples: When a person has a lung infection, fluid from the lungs is coughed up into a cup for collection.

TB-Positive: Having cells infected by tuberculosis.

TB-Negative: Having no cells infected by tuberculosis.

thesis: A large report based on research.

undergraduate: First four years of university or college study.

virologist: Someone who studies viruses, such as the flu.

Photo and/or Image Acknowledgements

Bibliography

APOPO, "A Message from Dr. Jane Goodall, APOPO's Friend and Advisory Board Member." YouTube video, 5:33. October 6, 2022, https://www.youtube.com/watch?v=tnRWgGmHMOo

APOPO, "Apopo Celebrates 25 Years of Life-Saving Achievements." August 2, 2023. https://apopo.org/latest/apopo-celebrates-25-years/.

APOPO, "Detecting Landmines." https://apopo.org/what-we-do/detecting-landmines-and-explosives/.

APOPO, "HeroRATs: APOPO's Unique Approach to Solving Global Challenges." https://apopo.org/jane-goodall-apopo-visit-2024/

APOPO, "How it works. APOPO's Tuberculosis detection Rats." YouTube video, 1:57. June 24, 2016, https://www.youtube.com/watch?v=Z_vc5BtPPQ0.

APOPO, "International Day of Women and Girls in Science," 2023. https://apopo.org/in-focus-dr-donna-kean.

APOPO, "Our History." https://apopo.org/who-we-are/our-history/.

APOPO, "Training HeroRATS for Search and Rescue." YouTube video, 2:31. May 16, 2022, https://www.youtube.com/watch?v=MHJk8ODT6es.

APOPO, "What does APOPO Mean." https://apopo.org/who-we-are/faqs/.

Centers for Disease Control, "What you Need to Know about Tuberculosis," October 2023. https://www.cdc.gov/tb/media/pdfs/What_You_Need_to_Know_About_TB.pdf.

CBC KIDS NEWS, "HeroRAT retires from sniffing out landmines." YouTube video, 1:39. July 8, 2021, https://www.cbc.ca/kidsnews/post/watch-this-rat-is-retiring-after-5-years-sniffing-out-landmines.

Cooper, Ross G.. The African Giant/Pouched rat (Cricetomys gambianus): it's Physiology, ecology, care and taming. Spain: Lulu.com, 2014.

Dionisio, Chloe. "Hero Rats are being trained to rescue earthquake survivors." Discovery, June 8, 2022, https://www.discovery.com/nature/rats-are-being-trained-to-rescue-earthquake-survivors-.

Eindhoven, "Philips' legacy in Eindhoven," This is Eindhoven, October 12, 2022. https://www.thisiseindhoven.com/en/city-life/about-eindhoven/philips-legacy-in-eindhoven.

Fox, Susan. RATS. Neptune City, New Jersey: T.E.H. Publications, Inc, 1988.

Frederick, Denver, "APOPO—Harnessing the Senses of Hero Rats for Humanitarian Progress," LinkedIn, July 10, 2023. https://www.linkedin.com/pulse/apopo-harnessing-senses-herorats-humanitarian-denver-frederick.

Gregory, Josh. RATS. Auburn, Alabama: Children's Press, 2016.

Grounds, Kate. (University of St. Andrews Research Manager), email correspondence with the author, August 19, 2024.

Kean, Donna., email correspondence with the author, 2022–2024.

Kean, Donna., SKYPE call with the author, August 19, 2022.

Kean, Donna, Research Gate, 2008-2024. https://www.researchgate.net/profile/Donna-Kean.

Living Links Research Centre, "What is Living Links?" 2024, https://living-links.org/what-is-living-links/.

Mackie, Tom (landscape photographer), email message to author, February 22, 2023.

McNicholas, June. Keeping Unusual Pets: Rats. Chicago, Illinois: Heinemann Library, 2010.

Miller, Olivia. (University of Kent Press & PR Officer), email correspondence with the author, November 13, 2024.

Parachini, Jodie. HeroRat!: Magawa, a Lifesaving Rodent. Chicago, Illinois: Albert Whitman & Company, 2022.

PDSA, "HeroRAT Magawa is awarded the PDSA Gold Medal." Youtube video, 13:02. September 25, 2020, https://www.pdsa.org.uk/what-we-do/animal-awards-programme/pdsa-gold-medal/magawa.

PhD Women Scotland (blog), "A day in the life of my PhD: when your participants are monkeys," Donna Kean, posted April 16, 2020.

PhD Women Scotland (blog), "Some poignant moments during a PhD and my mum's battle with cancer," Donna Kean posted December 16, 2021.

Reuters, "Rats with backpacks trained for search and rescue." Reuters video, 1:30. November 8, 2022, https://www.youtube.com/watch?v=C0p61tc-usE

Science, Forget dogs: These rats could be the future of search and rescue, Robin Donovan, December 16, 2021. https://www.science.org/content/article/forget-dogs-these-rats-could-be-future-search-and-rescue.

Shallom, Lily, email correspondence with the author, 2022–2024.

Shallom, Lily, Google meeting with the author, September 21, 2023.

Wetzel, Corryn, "Backpack-wearing rats could start search-and-rescue missions next year." New Scientist online article, June 17, 2022. https://www.newscientist.com/article/2324994-backpack-wearing-rats-could-start-search-and-rescue-missions-next-year/.

Acknowledgments

This book has been a labor of love, and I have many individuals to thank for its development and production.

Thank you to Dr. Donna E. Kean for sharing the story of her childhood and her work at APOPO. She reviewed the story, vetted the information, and shared photos included in this book. I so appreciate her kindness, candor, and willingness to share her story. She is inspirational in her work with animals.

Thank you to Lily Shallom, APOPO Communications Manager, who approved my proposal for the book so I could write about Dr. Kean and this amazing organization. She reviewed and vetted the APOPO information and assisted me with permission to use APOPO photos and a quotation from the website. I am very grateful for her time and support.

Thanks to the APOPO organization! Your amazing humanitarian work is very inspirational to me and many others. Wishing you much success in finding new ways to save lives and communities throughout the world.

Thank you to Erin Sackett and Leakena Hok, at the Siem Reap Visitor Center, for our Google Meet session, answering my questions and introducing me to Nina, the first African giant pouched rat I had the pleasure to meet.

Thank you to Margaret Gordon, the first reader of this book, as it evolved. Her support, input, and encouragement were immeasurable as this book came to life.

Thank you to several schools, their teachers and a librarian who invited me into their classrooms and library to share about Dr. Kean's childhood. Students, in grades 3-6, completed a survey identifying which areas of Dr. Kean's childhood they felt I should include in the story. The participating schools were Webster Montessori in Webster, New York (Teacher Alisa Sciolino); Charles Carroll School No. 46 in Rochester, New York (Librarian Caroline Keeler); and Our Lady of Mercy School in Rochester, New York (Teacher Katie LaDuke). I appreciate Elizabeth Guzzetta's assistance (Dean of Academics, Our Lady of Mercy School) in arranging the opportunity at Our Lady of Mercy School.

Thank you to my sons for their feedback on images, the Beta Reader agreement, and their support and interest in this book. I am grateful to my Beta Readers: Elizabeth Guzzetta; Joann Tetlow; Marsha Sherwood; Christina Wofford-Gaines; Christelle Roux, PhD, Principal Scientist; and librarian Andrea Snyder (Skaneateles Library). Their time and input for this biographical science book is much appreciated and validated my efforts in writing it. I appreciated the final review by Erica Regan, and librarians Caroline Wheeler, Elissa Schaeffer and Deena Viviani (Brighton Memorial Library).

Thank you to Riley S. and Sophia C. for their review of the story with their grandmas. I appreciate their feedback as well.

Thank you to the librarians and staff at the Brighton Memorial Library, Penfield Public Library, Pittsford Public Library and Fairport Public Library, the students in Amy Balcerek's classroom at Trinity Montessori School and my sons for their help in choosing the cover for this book.

Thank you to Ronda Roaring (https://www.rondaroaring communications.com) for her concise critique and copy editing prior to publication.

I appreciate the insight of artist Marie Stone-van Vuuren for her suggestions in the use of illustrations and photos in the book. Thank you to illustrator Sadie Peers for her beautiful illustrations in the front and back of the book. And an enormous thank you to Tanya Space, Strategic Graphic Designer, who brought this book to life with her heart and graphic design expertise. When I first saw her work, I knew my book would touch others' hearts as well.

J. N. Courtney, 2025

About the Author

J. N. Courtney loves to write about imaginative, innovative, creative ideas and people. She writes stories which she hopes will resonate with readers.

Dr. Donna Kean's childhood and her work with the HeroRATs at APOPO is an incredible story. Courtney believes so strongly in APOPO's work that a percentage of the royalties for this book will be given to APOPO. This is her first biographical science book.

Courtney has also written two children's fantasy/adventure books in the CLOUDSCAPE series: CLOUDSCAPE Charlie's Story and CLOUDSCAPE Matilda's Story. She is currently working on the third and final book in the series, CLOUDSCAPE Kagiso's Story.

Courtney retired from her work as a school-based Occupational Therapist for almost 26 years. When not writing at her home in upstate New York, she enjoys time with family and friends, museums, hiking, traveling, experimenting with recipes, reading, and gardening with the birds and squirrels in her backyard. You can contact J. N. Courtney via her website: *jncourtneypublications.com*.

Nine African giant pouched rats were chosen to be trained to become a RescueRAT (or a Rat with a Backpack). Seven special rats were able to complete all of the challenging training steps—Kiria, Jo, Pearce, Caruso, Billy, Daniel (named after Dr. Donna Kean's nephew), and Wagner (named after a researcher in animal training research).

DANIEL

WAGNER

BILLY

KIRIA

CARUSO

JO

PEARCE